FAMOUS AIRCRAFT OF THE
NATIONAL AIR AND SPACE MUSEUM

VOLUME 1
EXCALIBUR III
The Story of a P-51 Mustang

VOLUME 2
THE AERONCA C-2
The Story of the Flying Bathtub

FUTURE VOLUMES PLANNED
MESSERSCHMITT 262
LOCKHEED XP-80
ALBATROS DVa

2

NATIONAL AIR AND SPACE MUSEUM

Aeronca C-2
The Story of the Flying Bathtub

by Jay P. Spenser

PUBLISHED FOR THE
National Air and Space Museum
BY THE
Smithsonian Institution Press
WASHINGTON, D.C.
1978

© 1978 by the Smithsonian Institution. All rights reserved.

Printed in the United States of America

Designed by Gerard A. Valerio

LIBRARY OF CONGRESS CATALOGING IN PUBLICATION DATA
Spenser, Jay P.
   Aeronca C-2: the story of the flying bathtub.
   (Famous aircraft of the National Air and Space Museum; 2)
   1. Aeronca airplanes. I. National Air and Space
Museum. II. Title. III. Series.
TL686.A42S63   629.13'09   78-606098

ISBN 0-87474-879-8

FRONTISPIECE: The National Air and Space Museum's restored Aeronca C-2.

The paper used in this publication meets the minimum requirements
of the American National Standard for Permanence of Paper for
Printed Library Materials Z39.48-1984.

98 97 96 95 94 93 92 91 90 89

5 4 3 2

Unless otherwise noted, photographs are from the National Air and Space Museum files or belong to the author.

Contents

ACKNOWLEDGMENTS vii

I. Origins of the Aeronca C-2   3

II. Development of the Aeronca C-2 and C-3   15

III. Restoration of the First Production Aeronca   57

APPENDIXES

A. Specifications and Performance, Aeronca C-2   71

B. History of the First Production Aeronca, 1930-1940   72

Acknowledgments

The author wishes to thank those who made the writing of this book possible. Much information and material was graciously provided by many who share an interest in light aircraft, and the first Aeroncas in particular.

Special appreciation is due Paul R. Matt for making available much of his own research in the field. Mr. Matt reviewed the first draft of this book at great length, and his corrections and amplifications reveal a firsthand knowledge of the subject.

John Houser of Aeronca, Inc., provided a wealth of information and photographs. An engineer for the company, Mr. Houser's expertise made possible the accurate restoration of the aircraft, and his love of the plane resulted in the availability of much that might otherwise have been lost.

I would like to give special thanks to Peter M. Bowers for reviewing this work, and for providing many excellent photographs. Mr. Bowers is a noted aviation historian and author with a special interest in early general aviation aircraft. In addition, he is a longtime Aeronca owner with a seemingly limitless fund of detailed information.

Additionally, thanks are due to Florine Crockett of the FAA Aircraft Registration Branch; Richard Farrar, Smithsonian photographer, for his documentation of the Aeronca C-2's restoration; Robert C. Mikesh as curator in charge of the restoration; and Louise Heskett, editor, Smithsonian Institution Press. Also of substantial help were Walter J. Boyne, Don Dwiggins, Donald S. Lopez, and Kenneth R. Unger.

J. P. S.

Aeronca C-2

The Story of the Flying Bathtub

An early Aeronca C-2, photographed in 1930, reveals the comical lines that gave it the nickname "flying bathtub."

I

Origins of the Aeronca C-2

"Last year was epochal in the aircraft industry," the Aeronca advertisement of 1930 went. *"It saw the opening of the great private owner market through the introduction of the first practical light airplane . . . the now famous Aeronca C-2."*

The diminutive C-2 did indeed open an enormous, previously untapped market, winning enthusiastic acceptance during the latter half of 1930 despite the spreading depression. It was the first American airplane to be affordable, economical, and produced in quantity. In addition, it was easy to fly, required little maintenance for its simple structure or reliable engine, and was devoid of nasty habits to spring on the inexperienced pilot.

The advertisement quoted above is wrong only in that the Aeronca C-2 is not famous, for today it is all but forgotten. Neither sleek nor fast, it was not a plane to capture the imagination. It won no air races although it set a number of records, and was so small as to look like an overgrown model airplane.

The significance of the C-2 lies not in what it did but in what it was; this aircraft marks the emergence of general aviation in the United States. Like the keystone of an arch, its story is an integral part of the story of private flying. General aviation today is a major industry which ranges from fighting forest fires to helping businessmen operate more effectively. Private aircraft vary from single-seat homebuilt planes to multimillion-dollar jets. Cessnas, Pipers, Beechcraft, and others proliferate at airports across the country, and each year carry almost a hundred million people between cities in the United States.

Before the Aeronca C-2 came into being, manufacturers of civil aircraft catered only to "professionals"—a dubious term in the decade following the First World War—and "sportsman pilots." This latter category was made up of the independently wealthy and represented a very limited market at best. It was an era in which aircraft ownership invariably denoted high social status, an idea perpetuated by manufacturers who touted their products as the ideal way to attend yacht regattas and polo matches.

Flying for the average American was not possible with the large and expensive airplanes of the day. In Europe there were flying clubs, but an airplane owner in the United States had to undertake the costs of purchase, instruction, and maintenance himself. With the introduction of truly light aircraft, the Aeroncas and their competitors, this situation changed and flying schools sprang up everywhere.

The C-2 was a squat airplane with a fuselage shaped something like the bill of a pelican and a two-cylinder engine with a tiny propeller that made the plane look slightly bug-eyed. Balanced on top was a disproportionately long wing braced with wires that appeared woefully inadequate, as if the first

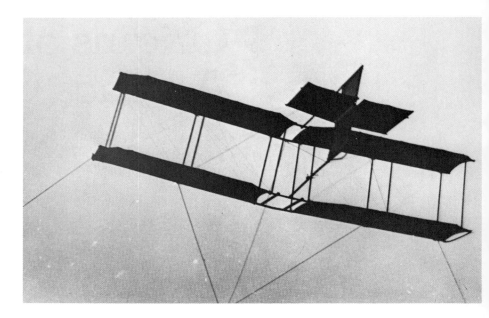

The Witteman Glider loaned to the New York Model Aero Club in 1911. (Courtesy Aeronca, Inc.)

gust of wind would topple it. The C-2, and later its big brother the two-seat C-3, were frequently called "flying bathtubs" or "Airknockers," and they were invariably greeted with laughter when first observed among the American Eagles, Wacos, and Travel Airs that dominated most airfields of the day. Although not heroic by any stretch of the imagination, the C-2 ushered in a new age and led a revolution that is still going on today.

The story of the development of the Aeronca C-2 began with Jean Roche, a French-born aircraft designer. Roche was twelve years old when his family arrived in America, and he developed a passionate love of aviation. A skilled model builder, he won awards for the excellent flying characteristics his models always possessed. He was a regular visitor at airfields on Long Island, New York, where there was a pioneer nucleus of aviators that included Moisant, Bellanca, Schmidt, and Burnelli. He learned with these men, observing their mistakes as well as their successes.

On occasional weekends, Roche went to Staten Island's Miller Field or Governor's Island where with other boys he learned to fly the Witteman hang glider. He was also a member of the New York Model Aero Club where his skill at laminating and carving small propellers earned him extra money.

When the Witteman glider used by the Model Aero Club was wrecked, Roche decided to build his own. It was foldable and designed to be carried to Van Corland Park from the Roche home at 102 West 90th Street on the New York subway system. Much to the relief of his mother, Jean had to abandon the project when he ran out of money to buy cotton with which to cover the wings.

It was at this time that Roche conceived a pivotal airfoil system which stabilized an aircraft in turbulent air. He applied for a patent on June 1, 1911, at the age of seventeen, and received it three years later. It was far ahead of its time, and aircraft companies showed no interest.

An additional experience with aviation came in the summer of 1910 when Roche was invited to test fly a "demoiselle" type monoplane built by a linotype operator named George W. Beatty. Starting at home plate of the then Yankee Stadium at 165th Street and Broadway, Roche opened the throttle and headed for the outfield where he ran into a ladder on the ground. The plane had attained a speed of about 20 miles per hour and

The New York Model Aero Club, about 1910. Jean A. Roche is second from left in the middle row. (Courtesy Aeronca, Inc.)

Demoiselle-type monoplane which Jean Roche attempted unsuccessfully to test fly in 1910. (Courtesy Aeronca, Inc.)

bounced into the air, flying a total distance of 10 feet at a height of 8 inches. While taxiing back, a fire broke out in the engine above the wing. Roche, soaked with gasoline, scrambled free as buckets of sand were dumped on the plane.

Already familiar with the rudiments of aviation science through association with the flying community and his own experimentation, Roche entered Columbia University where he graduated with a degree in mechanical engineering. During the college years, he kept a close watch on American and European aeronautical magazines, and made frequent visits to the flying fields around New York City.

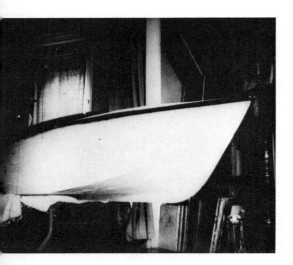

Hull of the canard flying boat project begun by Roche in his mother's apartment in 1914. The aircraft was never completed as Mrs. Roche ran out of patience and sold it. (Courtesy Aeronca, Inc.)

Roche began working in 1914 on a canard flying boat with a pusher engine. He built the hull in his mother's living room, filling the fifth-floor apartment with wood shavings until her patience failed and she sold the hull to a motorboat racing enthusiast. Such projects, however trying for Mrs. Roche, were excellent training for Jean's career.

In 1915, Roche went to work as a designer for the Huntington Aircraft Company. The following year, he moved to Toronto, Canada, where he designed the M.F.P. Airplane for the Polson Iron Works. His next job was as assistant chief engineer at the Standard Airplane Company in Plainfield, New Jersey, where he helped design the Standard J-1. Then in 1917, with World War I in full swing, he answered a call for aeronautical engineers from Captain Virginius E. Clark of the Army Signal Corps. Involved in the direction of the Army's aviation activities, Clark was gathering qualified men together to close the wide gap between Europe and the United States in aviation. Roche's title became chief of design and technical advisor to the Materiel Division at McCook Field, Dayton, Ohio.

What was to become the Aeronca C-2 was already forming in Jean Roche's mind. As quoted by Paul R. Matt in his work *Aeronca; Its Formation and First Aircraft,* Roche wrote in correspondence, "In those early days it appeared to me that flying's worst enemy was atmospheric turbulence; I early appreciated the deadliness of the stall and in seeking a remedy for both came upon the discovery of an automatically stable type aircraft. . . . This having been demonstrated in model form, I was determined to build and fly such a craft as soon as my humble means would permit."[1]

Roche's experience with both models and gliders convinced him that what was needed was a small and simple aircraft with excellent stability and the ability to glide efficiently. The growing prosperity of the 1920s was beginning to make itself felt and aircraft at that time were becoming bigger each year. To Roche, though, largeness for its own sake was simply poor engineering. Design techniques of the day were still largely improvisational, and the universal cure for unspectacular performance was to put in a larger engine.

The keynotes of Jean Roche's work were always sound engineering, simplicity, and functionality. Roche recognized that aircraft design is a series of compromises between such factors as weight, range, useful load, and fuel load. He kept weight to a minimum as a matter of course, and insisted upon strict functionality of parts.

A good chance for Roche to put his theories into practice presented itself when in 1923 McCook Field officials gave him the assignment of designing a glider for the Army Air Service. The resulting craft, designated the GL-2, was quite satisfactory despite being fitted with upper wing panels from a Curtiss JN-4 Jenny biplane as its single wing. General William Mitchell flew this glider at one point, and his comments to the designer were complimentary.

Another departure from established aeronautical practices was Roche's insistence that the monoplane was superior to the biplane. Roche was by no means the only proponent of the monoplane, but he was in a position to be heard. When his light plane flew, it outperformed the standard biplane trainers of the day with only a fraction of the power. He felt then that he had made his point.

Roche actually began work on his light plane in the early 1920s, although progress was very slow at first. Tests made in the wind tunnel at the Massa-

---

[1] Paul R. Matt, *Aeronca: Its Formation and First Aircraft.* Historical Aviation Album, vol. 10 (1971), p. 273.

GL-2 glider. (Courtesy Peter M. Bowers)

chusetts Institute of Technology on a new series of airfoils attracted his attention. One in particular, the Clark Y, appeared to offer excellent characteristics over a wide range of speed and attitudes. Roche chose this airfoil as his wing and with relief put aside the idea of making do with existing wings such as the Jenny panels used on the GL-2.

Despite the limitations of time and finances, the fortunes of the little airplane took a turn for the better. John Q. Dohse, an assistant of Roche's at McCook, volunteered his services as a craftsman and machinist. Dohse had a love of aviation dating from the time he saw Orville Wright fly one of the Wright Flyers, and he willingly gave up his free time to lend a hand.

With the airframe nearing completion, the major stumbling block became finding a suitable engine. There were small engines available in England and Europe, but they were for the most part too expensive to be considered. Unfortunately, there were no powerplants in the United States that matched the requirements of the airplane taking shape in Roche's garage at 28 Watts Street in Dayton, Ohio. To make matters worse, Roche sent an order to Germany for a 28-hp Haake engine, but his deposit was lost when the company went out of business.

Roche light plane fitted with 18-hp Henderson motorcycle engine, 1924. (Courtesy Aeronca, Inc.)

With the plane otherwise finished, Roche and Dohse installed an 18-hp Henderson motorcycle engine for preliminary tests. The engine proved insufficiently powerful to raise the small craft off the ground, however, which was just as well as the Henderson was subject to overheating and was extremely unreliable.

As fate would have it, a young engineer in the Powerplant Section at McCook had recently been asked if he would design a small, two-cylinder, horizontally opposed engine to pump air into the ballonets of a blimp. Harold Morehouse was interested in the possibilities of small gasoline engines, and readily agreed to come up with a preliminary design on his own time. Some months later, the first of his engines was completed, and in March 1924 it was installed in an airship. Roche happened to witness the tests.

Roche knew immediately that it was just the type of powerplant his little light plane needed. Rated at 15 hp, it was of course too small, but the overall configuration was correct. Morehouse in the Engine Department was a near neighbor of the Roche Design Group, and it was just a matter of hours before Morehouse accepted the challenge of designing a proper engine for the light plane.

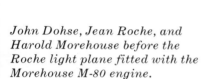

John Dohse, Jean Roche, and Harold Morehouse before the Roche light plane fitted with the Morehouse M-80 engine.

Jean A. Roche, 1925. (Courtesy Aeronca, Inc.)

The ballonet pump motor had a displacement of 42.2 cubic inches, with a 3-inch bore and 3-inch stroke. The airplane engine was to have a displacement of 80 cubic inches, a limit imposed by the National Air Race Committee for light planes and observed by Roche with an eye to the sporting uses he envisioned for the little plane. Working on his own time and machining parts on borrowed facilities, Morehouse completed the compact engine by the middle of 1925.

After the Morehouse engine had been bench tested approximately fifteen hours, the plane was disassembled and taken out to Wilbur Wright Field. A nonflying Barling bomber that was a fixture at the field became its new home, its broad wing sheltering the little aircraft almost as well as a hangar could have.

John Dohse became the first man to fly the Roche light plane when, on the evening of September 1, 1925, he added too much power during a taxi test. Although he had no previous flight experience other than taking the controls on occasion, Dohse decided to climb out and circle the field at a distance of five miles before returning to make a perfect landing. There was great jubilation among the three men, especially for Roche who had designed the plane to be safe and easy to fly for inexperienced pilots.

Harold Morehouse's engine was very successful as well. Rated at 29 hp, it perfectly complimented the airframe for which it had been designed. More than 200 flights were made during the next year, and it performed flawlessly. At one point, John Dohse took the light plane up more than 20,000 feet!

As quoted by Matt, Roche observed that his aircraft "was not designed for any particular type of competitive flying, but merely as a relatively safe, fair weather, economical airplane."[2] The praise of pilots who asked or were invited to fly it was unanimous, and the possibilities of owning such a plane suggested themselves. Many pilots wistfully pictured the small craft in a small hangar by the house, ready at a moment's notice to be wheeled out and

---

[2] Ibid., p. 275.

flown off the small grass field out back. It could be flown for the sheer pleasure of it, and purely by feel rather than by reference to instruments.

Two incidents occurred in 1926 that had a bearing on the future of the Roche light airplane, the immediate progenitor of the Aeronca C-2. The first was that Harold Morehouse was offered the opportunity to manufacture and market his engine with the Wright Aeronautical Corporation in Paterson, New Jersey. Roche, Dohse, and Morehouse agreed it would be for the best. Having the Wright-Morehouse engine available commercially would greatly improve the chances of Roche's plane also being produced, and in any event neither Roche nor Dohse wanted anything to stand in the way of the success of their friend.

The second incident was the first major accident of the Roche light plane. John Dohse was flying with his brother squeezed into the back when he attempted a steep bank at too low an airspeed. The overloaded plane stalled and crashed before Dohse could regain full control. He received a broken ankle and his brother was unhurt, but the front of the little airplane was destroyed and it was apparent the Morehouse engine would never run again.

Damage to the airframe was quickly repaired and Roche obtained the services of McCook Field engineers Roy Poole and Robert Galloway to design another engine. It was to be generally similar to the Morehouse engine, and where possible parts were to be traced and duplicated. The resulting engine was a bit larger, heavier, and less powerful, but it ran smoothly and offered a degree of reliability generally unknown in small aircraft powerplants.

The Poole–Galloway engine was to be known as the E-107, the displacement being 107 cubic inches. Modified somewhat to meet production requirements and redesignated the E-107A, it would be a key element in the success of the Aeronca C-2. It developed 26 hp at 2500 rpm, later increased to 30 hp before being superseded by the Aeronca E-113 developed for the two-seat C-3 aircraft of 1931.

But in the late 1920s, prospects for finding a builder for the light airplane Jean Roche had designed were bleak. Roche observed that big and cumbersome aircraft were selling, and they were bigger and more cumbersome than the year before. Another piece of bad news came when it was learned that the Wright Aeronautical Corporation had decided that there was not enough demand to justify producing the Wright-Morehouse engine any longer. In 1927, after selling only a few engines, they canceled the entire project.

Still believing his light airplane had a future as a production aircraft, Roche contacted the Govro-Nelson Company of Detroit, Michigan, and reached an agreement to have the E-107 produced if he could sell his airplane at any time in the future. Having done all he could, he put his light airplane aside, and he and John Dohse again devoted their full energies to their duties at McCook.

The next chapter in the story of the Aeronca C-2 started in November 1928 with the formation of the Aeronautical Corporation of America, a company as small as its name was big. Although it had assets of one-half million dollars, it had no airplane, designer, or even factory. The Board of Directors and Corporation Officers were J. M. Hitton, Hitton Brokerage Firm; I. C. Keller, Dow Chemical and Drug Company; J. M. Richardson, Richardson Paper Company; Taylor Stanley, American Laundry Machinery Company; and Ohio's senior senator, Robert A. Taft, son of the former president.

This diverse group had no idea that theirs would be, as stated in the 1931 edition of *Jane's All the World's Aircraft*,[3] "the first American company to

The Metal Aircraft Company's "Flamingo."

build and market a truly light aeroplane." First things came first and finding a location for Aeronca, a contraction of the company name, seemed the logical starting place. The board members chose the newly constructed Lunken Airport near Cincinnati, Ohio.

Conrad Dietz, a businessman and aviation promotor in the area, caught wind of Aeronca's unusual situation and offered the company a biplane—the C-4—for which he had acquired the rights, along with a partially completed prototype and detailed construction methods. Attractive as the offer sounded, however, by now the monoplane appeared to be the wave of the future and the board was not overly enthusiastic.

A pilot himself, Dietz recognized the limited production life his C-4 would probably enjoy. He decided on a different approach and suggested they take a look at the light monoplane his friend Jean Roche had designed and built. The fact that Roche was senior aeronautical engineer for the United States Army Air Service was in itself enough of a recommendation and the board members agreed.

Arrangements were made with Roche for a demonstration flight, and he and John Dohse hurried to put the light plane in top condition. When the directors of Aeronca were assembled at Wright Field, John Dohse climbed into the plane and lifted it off the ground to put it through its paces overhead. It was a flawless performance and the Poole–Galloway engine never missed a beat.

[3]*Jane's All the World's Aircraft, 1931*. Compiled and edited by C. G. Grey and Leonard Bridgman (London: Sampson, Low, Marston & Company, Ltd., 1931) p. 242c.

While Dohse flew, Roche explained the strength and simplicity of the steel-tube fuselage and wooden wings. He also stressed the safety inherent in its 11-to-1 glide ratio and very low landing speed of 30 miles per hour. If a total engine failure was experienced while flying 1 mile above the ground, he explained, the pilot would be able to glide more than 10 miles before landing, and the low landing speed meant that only a few hundred feet of room were required to roll to a stop.

The assemblage was clearly impressed, but were guarded when they approached Roche about the possibility of having his light airplane manufactured commercially. Several weeks later, Roche received a firm offer and was ecstatic. In the meantime, he had been making last-minute improvements and performing engineering tests of all kinds on his plane.

John Dohse left in April 1929 to work for the Boeing Company in Seattle, Washington, and once again Roche was on his own; however, Aeronca now had a paternal interest in the well-being of their product-to-be and was attending to many of the details of getting the ball rolling. For Roche, it was a very happy period.

Aeronca turned its attention first to working out the terms of production of the E-107A engine by Govro-Nelson. All parts would be manufactured in Detroit and shipped to Lunken Field for assembly, testing, and installation in the light planes. There was still not a factory, but Aeronca had its eye on the Metal Aircraft Company and its spacious plant where the "Flamingo" cabin monoplane was built. The company was not doing well and some floor space might be available for the building of Aeroncas.

On June 8, 1929, Roche's light plane was extensively tested by Maj. Gerald Brower at McCook in a flight which lasted 3½ hours. Brower, as described by Paul R. Matt, "pronounced the plane virtually faultless. Stability was exceptional and turns could be made with stick and rudder, rudder alone or stick alone with no slipping. The ailerons gave a slight incorrect yaw and he suggested they be rigged to trail high. . . . Top speed was 80 mph, stalling speed 35 mph and climb at 700 lbs. gross weight was 5,000 feet in 13 minutes. The plane would spin readily below 30 mph IAS, but would recover in 1½ turns once all controls were turned loose."[4]

In addition, Major Brower suggested that brakes and top windows be added, and that the power be increased slightly. He wrote his impressions and gave the report to Roche to present to Aeronca when the plane was officially handed over.

The deal was finalized early in the summer of 1929. In return for all rights to manufacture the aircraft, Roche was given 220 shares of Aeronca stock. Roche did not want to leave his position at McCook, and it was arranged that he have an advisory status rather than be Aeronca's aeronautical engineer. Robert B. Galloway, the McCook Field powerplant engineer who had helped redevelop the Morehouse engine, would later become Aeronca's first official chief engineer.

Meanwhile, Aeronca hired Roger E. Schlemmer of the University of Cincinnati's Aeronautical School to reengineer the plane as it was totally unsuitable for factory production methods. In the process, he added a small winddhield and top windows. Major Brower's other suggestions, brakes and more power, were added later in the course of C-2 and C-3 production.

Meanwhile, the Metal Aircraft Company was in serious trouble. The twentieth Flamingo had been built, but without a buyer for it they did not have

[4]Matt, *Aeronca*, p.277.

Employees of the Metal Aircraft Company display their faith in the "Flamingo." Fifty-eight men and full fuel in the tanks put more than 10,000 pounds on the wings. However, the plane was not economical and this example was the last one produced.

the means to build any more. Their plane was a seven-passenger cabin monoplane, with corrugated metal wings and fuselage that was simply too expensive and uneconomical to be competitive, and they were teetering on the edge of bankruptcy. For the people at Aeronca, it was a sobering reminder of the precarious nature of the airplane business.

Metal Aircraft readily agreed to let Aeronca rent space in their factory. With the fall of the stock market in October 1929, they went out of business officially and Aeronca bought the factory, tools, and equipment for a small fraction of their actual worth. If there was ever a time to build small and economical airplanes, this seemed to be it.

Aeronca NX626N, serial #2, on its maiden flight in the fall of 1929. This was the first airplane built by Aeronca.

The Aeronca C-2 was a delight to fly. Built primarily as a sport plane, the cockpit was open, and the pilot flew by feel as there were few instruments.

The C-2 was actually a powered glider whose long wings gave it a shallow gliding angle and a low landing speed.

II

Development of the Aeronca C-2 and C-3

The first C-2 produced by Aeronca made its maiden flight on October 20, 1929. It bore the registration number X626N, serial number 2, and was painted bright yellow and orange with a rakish stripe running the length of the fuselage. This aircraft is now in the National Aeronautical Collection of the Smithsonian Institution. Its history and restoration are described later in this book.

Roche's light plane was given serial number 1 and marked "Aeronca C-2" in bold letters along the fuselage, but it was not a true Aeronca as it had been hand built several years before the formation of the company.

The second production Aeronca, predictably registered NX627N, flew nine days after the first. (The First "N" denotes United States registry, but was deleted from rudder and wing markings). Together, these two planes were rushed to the Western Aircraft Show held November 9-17, 1929, in Los Angeles, California. It was the public unveiling of the Aeroncas, and they caused a sensation. The novelty of owning a private airplane for less than $1,500 fascinated visitors and made a lasting impression on the aviation community.

The man responsible for bringing the C-2 to the public's attention was A. J. Edwards, sales manager for "Aeronca". Edwards had previously been associated with the Ryan Aeronautical Company at the time of the building of the *Spirit of St. Louis* for Charles A. Lindbergh. A dynamic man with a genuine talent for salesmanship, Edwards realized that aircraft ownership would be considered a risky investment at best and an absurd extravagance at worst in the troubled economic climate which faced the nation. Nevertheless, he believed that the Aeronca C-2 would sell itself if given the exposure it needed.

Edwards next took the two Aeroncas to the International Aircraft Exposition and the National Conference on Aeronautics Education, held concurrently in February 1930 at Saint Louis, Missouri. "Iron Hat" Johnson, a skilled airshow performer, flew one of the Aeroncas and fell in love with it. Johnson, whose real name is Forest M. Johnston, began a lasting acquaintance with the C-2 and was in large measure responsible for its success as will be described later.

Another noted aviation figure to fly the Aeronca C-2 was Jimmy Doolittle. When he expressed an interest in trying out the new plane to "Iron Hat," the latter promptly made the necessary arrangements. Thinking back to the occasion, Doolittle recalls that the C-2 was "an excellent little airplane."

The Roche light plane at the Aeronca factory in 1929.

Edwards continued to take the Aeroncas from exposition to exposition around the country. Although somewhat ignored by the aviation publications of the day, they were enthusiastically received everywhere they went. With a base price of $1,495 at the factory, reduced by June 1931 to $1,245 as a result of the Great Depression, there was certainly no question as to the competitiveness of the C-2. In addition, few if any manufacturers could match the cost for both gasoline and oil of just one cent per mile.

Others were not slow to take up the challenge, however. The day of the light airplane had arrived and the marketplace was soon to be flooded. Some of the competing aircraft were extant designs that for various reasons had not enjoyed the Aeronca's success. Others were hurriedly produced, and generally lacked the advantage of designers with Jean Roche and Roger Schlemmer's credentials.

An Aeronca advertisement from 1931 refers derisively to the "scores of light planes being hurried on to the market." While this statement is a slight exaggeration, the aviation community was indeed deluged with what was termed the "flivver-plane movement." They numbered twenty-three, and with their appearance they collectively took from Aeronca the absolute dominance of the field which it had held all through 1930. A brief description of each of these light aircraft is given below to provide a better understanding of the times.

Famous race pilot Roscoe Turner and mechanic Don Young tried out the brand new Aeronca C-2 at the Western Aircraft Show in 1929. They appear to have enjoyed the experience, despite the absence of a second seat for Young to sit in. The C-2 had no brakes, but could be effectively stopped by a pilot wearing gloves.

First in alphabetical order was the Akerman Pusher JDA-8, an experimental pusher monoplane with two side-by-side seats and a triangular boom tail. The designer was Professor John D. Akerman, head of the Aeronautical Engineering Department of the University of Minnesota.

Next was the Alexander "Flyabout," D-1. With sales of its hardy "Eaglerock" biplane failing, Alexander Aircraft in Colorado unveiled the little Flyabout at the Detroit Air Show in early 1931. It was powered by the new 37 hp Continental A-40 engine, soon replaced by the 45 hp Szekely SR-3-0 in the D-2 model. Although an excellent airplane in all respects, offering such advanced features as a fully enclosed cabin and struts instead of wire bracing, the Flyabout never really caught on.

The American Eaglet.

The American Eaglet was a junior brother to the company's successful "Eagle" series of aircraft. Powered either by a 25 hp Cleone or 30 hp Szekely engine, it was a parasol monoplane with tandem seating for two in open cockpits. It was successfully marketed as a trainer.

The American Sunbeam Pup was a little-known two-seat light plane with a 40-hp Salmson engine, a 95-mph top speed, and a base price of $1,450. Application was made to the Department of Commerce for approved type certification, but the plane did not enter production.

Next in the list is the Arrowhead Safety Plane B-2, built by the Safety Aircraft Corporation of Miami, Florida. Priced at $1,500, this radical design featured wings swept back 30 degrees with all control surfaces located at the wingtips. How safe the Arrowhead Safety Plane was is open to conjecture, for it was never produced in any number.

A more successful plane was the Buhl "Bull Pup," a mid-wing monoplane that offered good performance and was well liked. The Bull Pup combined a conventional wooden wing with an advanced metal fuselage, and sold for $1,250. It was powered by a 3-cylinder Szekely radial engine of 45 hp.

The Cessna EC-2 offered a very clean design and bore a greater resemblance to later Cessnas than did its bigger brothers. Called the Baby Cessna, it featured an advanced, full-cantilever, tapered high wing. It was powered by an Aeronca engine, an unusual occurrence as these aircraft were competitors in the light-plane market.

The American Sunbeam Pup LP-1.

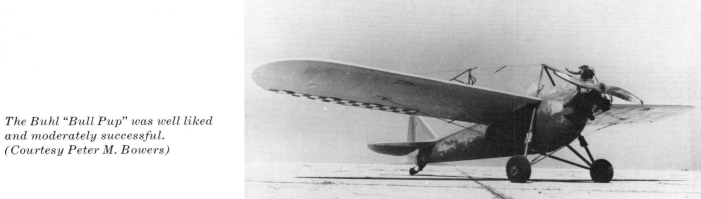

The Buhl "Bull Pup" was well liked and moderately successful. (Courtesy Peter M. Bowers)

The Cessna EC-2 was powered by an Aeronca engine and had an advanced wing without struts or wire bracing.

A Curtiss-Wright CW-1 "Junior" lands in a snowy field. Cockpits were open and required warm clothing in the winter.

The Church "Mid-Wing" was an ungainly single-seater powered by a 27 hp converted Henderson motorcycle engine under an automobile-style hood. A novel feature in the design of the Mid-Wing was the presence of a window in each wing to give the pilot some view downward and forward, especially necessary as the high hood blocked all view forward even in level flight! The Mid-Wing was advertised as being available at the Chicago factory for just $890, and plans were offered for the buyer who wanted to build his own.

Certainly one of the most significant aircraft to emerge from the flivver-plane movement was the Curtiss-Wright CW-1 "Junior," an example of which is included in the National Air and Space Museum of the Smithsonian Institution. The Junior was a two-seat open-cockpit parasol monoplane with a 45-hp Szekely engine mounted facing rearward above the wing. Offering excellent visibility and a top speed of 80 mph, it introduced many novices to the joys of flight. More than 270 were built before the depression put a halt to production in 1932.

Tenth in alphabetical order is the Cycloplane Solo and the Ground Trainer, an unusual "packaged" light-airplane instruction system featuring a single-place parasol-wing open monoplane with a 22-hp Cyclomotor A-2-25 two-stroke engine set into the wing leading edge. Instruction was given in a trainer capable of flying only slightly in ground effect, after which the student was turned out in the Cycloplane Solo. Included in the $1,990 price tag were text books, two flight suits, and a replacement engine.

The Cycloplane Ground Trainer. (Courtesy Peter M. Bowers)

Cycloplane Solo Trainer.

The Eyerly Coupe was a two-seat, side-by-side, cabin monoplane with the control stick overhead. An unsuccessful airplane produced by the Eyerly Aircraft Corporation of Salem, Oregon, it was redesigned and rebuilt with equally unspectacular results. Power was derived from a 37-hp Continental A-40 engine.

A moderately successful light airplane was the Heath Parasol, also offered as the Long-Wing and Center-Wing with different wing configurations. Wheel brakes and electric starters were available, and all three models sold for $975 in kit form.

The thirteenth light airplane was the Hunt Sport Chummy, a forgotten two-place cabin biplane design undertaken by the Hunt Aircraft Manufacturing Company of Fiskeville, Rhode Island. It was to be powered by a 37-hp Continental A-40 engine and sell for $1,650.

The Irwin Meteorplane FA-1 was a true light airplane in terms of size and power, but otherwise bore more resemblance to a World War I fighter than its contemporaries. An attractive single-seat biplane of 90-percent wooden construction, it was powered by a 4-cylinder Irwin 79 radial engine rated at 25 hp. Top speed was listed as 85 mph.

The Eyerly Coupe, or "Whiffle Hen" as named by the Eyerly Aircraft Corporation of Salem, Oregon.

The first Heath Parasol was hand built by two mechanics near Saint Louis, Missouri. Here, five-year old Leonard Lieurence, who lived near the airport and whose ambition was to fly around the world, poses in the Parasol.

The Irwin Meteorplane FA-1 predated the Aeronca. Looking like a miniature World War I fighter, it was built purely for sport.

Mattley Fliver No. 1.

Another hopeful design was the Lee Monoplane L-2 by the Lee Aircraft Company of Napoleon, Ohio. Powered by a Continental A-40 engine, it was a two-place open monoplane with an optimistically estimated high speed of 85 mph. It featured an adjustable stabilizer for trimming the plane in flight. A prototype was built and began flight tests, but the L-2 was not produced.

Sixteenth is the Lightplane Paraquet 5-M, a single-place cabin monoplane with a 25-hp Cleone engine. Several different models of this type were ready to be flown away from the factory, and plans were available so that the plane could be built by the buyer if so desired. Top speed was 70 mph.

An ungainly single-seat monoplane with a semi-enclosed cabin was the Mattley "Fliver" No. 1. It had a very thick wing topping a bulky fuselage, and was designed to accommodate engines between 25 and 60 hp. Top speed of a Continental A-40 powered Fliver was a rather modest 70 mph, and the price was $1,290.

The Nicholas-Beazley NB-8 light airplane featured side-by-side seating for two and a parasol wing that folded. It was fully aerobatic, and quite stable due to its long fuselage. Power was supplied by a 45-hp Szekely engine and the top speed was 85 mph, raised the following year to 110 mph by the addition of a NACA cowling and a more powerful engine.

The Peters Play Plane was a single-place open monoplane intended for production by the Peters Aircraft and Engine Company of Albany, California. Price was announced as $985, with an optional two-seat version to sell for $1,185.

Next in order is the Pietenpol B-4A two-place parasol light plane powered by a converted Ford Model-A engine rated at 40 hp. Construction was totally conventional and the top speed was 90 mph. Price at the H. B. Pietenpol factory in Spring Valley, Minnesota, was $850.

The Prest Baby Pursuit was a single-place high-wing monoplane built expressly for high performance by Prest Airplanes and Motors of Arlington, California, for the "lone pilot who wants to go places and do things." Powered initially by a 45-hp Szekely engine, quickly replaced with a 60-hp Lawrance, the top speed was 115 mph and it was advertised that the Baby Pursuit held the world's speed record for light airplanes. The fuselage was constructed of

Nicholas-Beazley NB-8.

A Pietenpol B-4A, photographed at an airshow.

Taylor E-2 Cub, forerunner of the Piper J-3 Cub.

steel tubing in a diamond cross section. It sold for $1,675 to $2,000 depending on how it was equipped.

Equally expensive was the Simplex Kite at $1,995. An experimentally built flying-wing design built by the Simplex Aircraft Corporation of Defiance, Ohio, it was a two-place low-wing aircraft powered by a 40-hp Szekely engine. In contrast to the Baby Pursuit, the Kite had a top speed of only 75 mph.

The Taylor Cub E-2 was a tandem two-place high-wing monoplane powered by a 37-hp Continental A-40 engine. Top speed was 75 mph, cruise speed was 62 mph, and landing speed was a gentle 26 mph. The price at the Taylor Aircraft Company in Bradford, Pennsylvania, was $1,325. The Taylor Cub would evolve into the world-famous Piper J-3 Cub, a plane whose name is synonymous with general aviation and light aircraft.

These light airplanes appearing in 1931 were not all the types to be developed, but rather just the first wave of a trend that would continue through the 1930s. These were airplanes built with no other intended purpose than to be inexpensive and fun to fly. World War II put a halt to the light plane movement, and it emerged again only slowly despite predictions of a private airplane boom spurred by the great number of returning military pilots.

An interesting sidelight to the story of the Aeronca C-2 is that the development of Jean Roche's light airplane is today paralleled by members of the Experimental Aircraft Association. Members build their own airplanes for several reasons, among them the desire to own a plane that is both less expensive and more fun than the products of the major manufacturers. With these requirements in mind, it is not altogether surprising that among the kits available to the home-builder are a number that bear strong resemblances to such planes as the Aeroncas and Curtiss-Wright "Juniors."

By 1931, most light aircraft had two seats rather than one. Airplane owners naturally wanted to share the pleasures of flight, and even more important was the fact that a second seat was necessary if the plane was to be used for flight instruction. In order not to lose its lead in the field, the Aeronautical Corporation of America set about developing the C-3, a larger version of the C-2 with seating for two side by side. Power was to be provided by the new Aeronca E-113 engine of 36 hp. Two prototype C-3s were put

Prest Baby Pursuit.

The Piper J-3 Cub, although not one of the first generation of light aircraft, is the plane most often associated with the beginnings of general aviation.

An early Aeronca C-3. (Courtesy Don Dwiggins)

through extensive tests in 1930 and production started in March of the following year.

One C-3 was dispatched on a 13,000-mile demonstration tour through seventeen states, and gave many people their first look at the "Duplex" or "Collegian" as the regular and trainer versions were called. Despite the added fuselage size and weight, performance was crisp with the new engine. Orders poured in and production literally had to double at Lunken Airport to meet the demand.

The 70-foot takeoff roll, 12,000-foot ceiling, and landing speed of only 35 miles per hour were good selling points, but the strongest was its economy combined with a utility the C-2 had not had. A good demonstration of its ruggedness was provided when a C-3 was entered in the 1931 National Air Tour, a difficult 4,858-mile event in which the new Aeronca, the first light airplane ever to participate, averaged a respectable 64 mph.

Aeronca owners all over the country flew for the pure pleasure of it, most after less than five hours of instruction due to the straightforward characteristics of the aircraft. The C-2 and C-3 were basically powered gliders with excellent gliding ability and gentle landing speeds. In addition, the pilot sat so low in the cockpit that the term "seat of the pants" applied perhaps better to landing an Aeronca than flying it. The result was that it was extremely difficult to make a bad landing as the pilot had an excellent view of the proximity of the wheels to the runway.

But the nimble Aeroncas could also perform brilliantly in the hands of a truly fine pilot, as was demonstrated many times over by Forest M. Johnston. Johnston had learned to fly as a cadet at March Field, California. A natural pilot, his career ranges from the earliest biplanes to jets, and he has been an inventor, a pilot for two presidents, airline owner, record setter, movie stunt pilot, and test pilot. But he is perhaps best known as "Iron Hat"

Owner Fred Haight with his early C-3, serial #A-107, at Lunken Airport.

Aviation author and historian Peter M. Bowers bought an Aeronca C-3 for $200, licensed and flyable but without a propeller, in 1951. Note that the tail skid has been replaced with a wheel. (Courtesy Peter M. Bowers)

The pure enjoyment of flight is captured in this view of Bowers flying his Aeronca C-3. It is the oldest two-seat Aeronca currently flying. (Courtesy Peter M. Bowers)

This very early Aeronca C-2, the eighth production Aeronca, anticipates the use of private planes as business aircraft.

Landings were easy to judge as the pilot had an excellent view of the proximity of his wheels to the ground, although the proper technique was to look halfway down the runway.

Johnson (omitting the "t"), a top-notch airshow performer who did in little airplanes what most of his peers would have hesitated to attempt in anything less than a specially stressed aerobatic plane.

"Iron Hat," so called because of the ubiquitous derby which became his trademark, was working for the Nicholas–Beazley Airplane Company, Inc., of Marshall, Missouri, at the time he first flew the Aeronca. A firm believer that little airplanes would do the most for the greatest number of people in aviation, he agreed to take over marketing and publicizing the Aeronca C-2 on the West Coast.

He accomplished his objective by becoming a featured attraction at airshows in his Aeronca C-2 NC-568V, serial #18. Billed as the "world's greatest light plane ace" and often compared to Ernst Udet, "Iron Hat" performed loops, spirals, slips, and other aerobatic maneuvers, handkerchief pickups with his wingtip, and more. An intelligent performer, he capitalized on the comical lines of his plane and always flew with his derby on. The combination was an image that was unmistakable among the ranks of airshow pilots.

"Iron Hat" would climax a performance by shutting off the engine and doing additional maneuvers before landing "dead stick" before the crowd. He carried it further at times, however; due to the small size and open construction of the Aeronca C-2, he could restart the engine in flight by climbing half out and spinning the prop by hand.

Moreover, Johnston made excellent use of the plane's ability to fly at very slow speeds. He became the first pilot to pick up mail from a man running along on the ground, he took off from the top of a speeding car, and he became the first man to refuel in flight from a boat.

In November 1930 at an airshow in Alameda, California, "Iron Hat" extended a cane held by its base from the cockpit of his Aeronca and hooked the handle of a can of gasoline being held out by a man standing on the rear

"Iron Hat" refuels in flight at an airshow in Alameda, California, in November 1930. (Courtesy Don Dwiggins)

Wearing his ubiquitous derby hat, airshow performer "Iron Hat" Johnson, whose real name is Forest M. Johnston, prepares to take off from atop a speeding automobile in his Aeronca C-2. (Courtesy Don Dwiggins)

Aeronca C-2 (NC568V, serial #18, in which Johnston performed at airshows and set several records.

The Aeronca "Collegian" offered a completely enclosed cabin with door panels that were removable for summer operation.

bumper of a fast-moving Austin automobile. An amazed crowd watched as the can was drawn into the plane where Johnston refueled in flight. It was the first time this had ever been done, and it served to demonstrate the stability and ease of flying of the little C-2. If Aeroncas were selling, a good portion of the credit must go to Johnston.

Ninety Aeronca C-2 aircraft were sold in 1930, an impressive number by pre-World War II standards, and seventy-four more were sold in 1931 despite the introduction of the two-seat C-3 and the appearance of other light planes on the market. Nevertheless, the end of the single-seat Aeronca was drawing near, especially after the loss of a substantial number of completed and partially completed C-2s in a storage-area fire. This fire seemed to clean the slate for a fresh start with the C-3, however, and optimism abounded at the Aeronca factory in Ohio.

In 1931, the C-2 became the "Scout" and the C-3 became the "Collegian" or "Duplex," the former a trainer and the latter a more expensively appointed private machine. In addition, the young and vigorous Aeronca Company developed two other versions.

The first was the Aeronca C-1 "Cadet," a radical departure from the C-2 which it so closely resembled. A detailed examination would reveal that 3½ feet had been clipped from each wing so that the wing area was reduced by 30 square feet. The rate of roll was improved, but the landing speed jumped to 40 miles per hour. The fuselage was built up of heavier gauge tubing and the dihedral of the wing was reduced to 2 degrees. Wing structures were strengthened, and all bracing wires were made of a heavier gauge so that the plane was capable of withstanding 8.5 Gs.

The C-1 was to be the Aeronca for the private owner who wanted a plane he could throw around the sky in any aerobatic maneuver without fear of overstressing his plane. The paint scheme was a rich combination of maroon and bright vermillion in a two-tone pattern, with careful attention paid to interior appointments and passenger comforts.

Aeronca C-1.

To provide the extra power required for competition and aerobatics, it was planned to install the Aeronca E-113 or E-113A engine developed for the C-3. Although providing only six to ten more horsepower than the E-107A, this installation turned out to be well worth the effort as it made the Cadet a very zippy airplane.

Conrad G. Dietz, the promoter who had brought together Roche's light plane and Aeronca, and now vice president and general manager of the company, took a personal delight in putting this new plane through its paces. His exuberance bordered on recklessness but the plane—NX-11290, serial #A-122—showed no external signs of stress. However, on September 12, 1931, Dietz was demonstrating the C-1 at Sharonville, Ohio, when the wings folded and he plunged to his death.

Just two weeks before, the Aeronca C-1 Cadet had received its type certification and production had been given the go-ahead. With Dietz' death the project lost impetus, and what would undoubtedly have been a very popular airplane among sporting pilots was never built beyond the first four. Of the surviving clipped-wing Aeroncas, one was converted to a C-2N and the other two were reportedly crated and put in storage. Price for the C-1 was listed variously as $1,595 and $1,695; with floats to make it a seaplane, it would have been $2,095.

The second new Aeronca was designated the C-2N "DeLuxe Scout." It was virtually indistinguishable from the standard C-2, but was equipped with the 36-hp E-113 engine for better performance. Unlike the spartan C-2 with its plywood seat and minimal attention to pilot comfort, the C-2N had deep cushions and a fuselage 6 inches wider for more elbow room. Behind the pilot was a 50-pound-capacity baggage compartment, or an optional 5-gallon auxiliary fuel tank could be installed. With a fuel consumption of 2.5 gallons per hour for the E-113, the auxiliary tank provided a full two hours of additional cruising time.

Attention to detail and appointment was greater even than with the C-1.

An Aeronca C-2N, featuring a C-3 rudder and mounted on Edo floats.

The interior was neat and the cockpit was now semi-enclosed with the addition of a wraparound windscreen. This windscreen and a tripodal landing gear were standard on all Aeroncas by 1932. Several paint schemes were offered, the standard one being an attractive two-tone blue with yellow trim for the fuselage and bright orange for the wings. Price for the DeLuxe Scout was $1,695, substantially more than the $1,245 then being asked for the standard C-2.

Although offered well into 1933, only four C-2Ns were built (two being converted from standard C-2s and a third being the converted C-1). Production of the basic C-2 had for all intents and purposes been halted by 1932, although Aeronca offered them on special order for another two years.

The C-3 was now Aeronca's star performer. As it could be mastered by the novice in two to four hours of dual instruction, many flying schools would give preliminary training in the C-3 before sending their students off on their own in the even more economical C-2. Soloing cost about $35 in 1931, and getting a license just twice that amount. Even at these rates, flying schools could make a healthy profit. With growing enrollments, many schools went out of the red while enjoying the luxury of operating planes that were both safer and easier to maintain. The Curtiss JN-4 Jenny may have been inexpensive to buy in the decade after the First World War, but it cost three times as much to fly and required a great deal of maintenance. The Standard J-1 was much the same.

The Aeronca E-113 engine was the most powerful 2-cylinder aircraft engine of its day. A horizontally opposed powerplant with a displacement of 113.5 cubic inches, it developed 36 to 40 hp and had a compression ratio of 5.1 to 1. The bore was 4¼ inches and the stroke was 4 inches. The entire engine,

The economy of the C-2 is well illustrated in this photo of an early model offered for rental at $4 per hour. (Courtesy Peter M. Bowers)

The instrument panel of the C-3 was mounted beneath the padded rear end of the fuel tank. Rudder pedals were close set and small, and a pilot flying solo generally straddled the single stick and used the outboard pedals.

The tripodal landing gear used on earlier C-3s is well illustrated in this view of Peter M. Bowers in his Aeronca. (Courtesy Peter M. Bowers)

An Aeronca C-2 on early-style floats. (Courtesy Don Dwiggins)

Aeronca C-2s and early C-3s were known as "razorback" airplanes, a term well illustrated in this view of a PC-2 taking off from a lake.

dry and minus the propeller hub, weighed 113 pounds. It ran with reassuring smoothness and was quite reliable, although it was subject to crankshaft failures. Operating cost per hour was just seventy cents.

Aeronca kept its finger on the pulse of the aviation public throughout the production of the C-2 and C-3. Improvements and desired features were quick to appear, the most noticeable being the new tail group and the extended, split-axle landing gear with Goodyear 16 x 7.3 "airwheels." Brakes were also quickly introduced as an option to facilitate ground handling.

Aeroncas were also offered as seaplanes after the first fifty C-2s had rolled out the factory door. The single and two-seat versions were designated the PC-2 and PC-3, respectively, the "P" standing for pontoon. Floats were manufactured by Warner; Aircraft Products Corporation; and the Edo Corporation of Long Island, New York. Edo was the largest supplier and also offered their Model D-990 and D-1070 floats as a conversion kit for wheeled Aeroncas at a price of $750.

Aeronca PC-2.

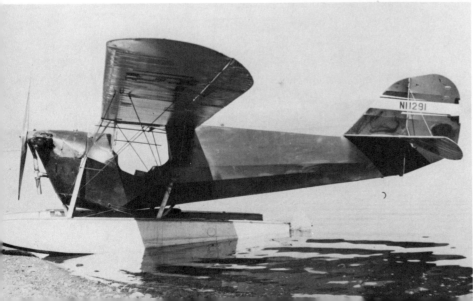

The Aeronca PC-3. (Courtesy Peter M. Bowers)

That the Aeronca C-3 was becoming a sophisticated aircraft is evident in this view. Unlike the sporting C-2, this C-3 with its enclosed cockpit has a decidedly utilitarian look. The pitot tube atop the cabane strut indicates that this C-3 has an airspeed indicator.

The Aeronca C-3 for 1935 featured a new landing gear with lower aerodynamic drag.

Although the internal structure remained unchanged, a wooden framework rounded out the fuselage lines of the 1935 Aeronca C-3. The "roundback" Aeronca also sported a new tail.

Compared to its predecessors, this late-model Aeronca C-3 was an aerodynamically very clean airplane.

The year 1932 saw the introduction of detachable cockpit doors with inset windows. These panels could be put on for winter- or bad-weather flying, or left off to enjoy warm summer weather to the fullest. The following year, the seat back was raised, the cockpit was widened slightly, and leg room was increased. The improved Aeronca E-113A engine was put on a new mount to reduce the level of noise and vibration experienced in the cockpit. In 1934, a new cantilever landing gear without drag-inducing struts became part of the C-3 design, and the interior was made more elegant with new leather seat cushions. With wheel pants, available as an option since 1931, the two-seat light plane attained a top speed of a little over 80 mph despite its low horsepower.

The looks of the C-3 changed dramatically in 1935 with the introduction of the "roundback" C-3 "Master." The triangular fuselage frame was retained, but its distinctive razorback appearance was eliminated with the use of a nonstructural wooden frame. This change filled out the shape of the fuselage and improved the airflow over the tail surfaces. As a result, a new smaller vertical stabilizer and rudder were adopted that made the C-3 look distinctly roundtailed. This was one of the very few cases of a control surface and stabilizer being reduced in area during the course of the development of an aircraft.

In addition to these changes, the 1935 Aeronca came with a completely enclosed cockpit, offered wheel brakes and wing lights as optional equipment, and sold for just $1,890. When compared with the $1,880 price originally announced for the C-2 before the Great Depression, the Master was a bargain. Production was doubled again, and the five hundredth Aeronca was turned out in 1935 with 128 C-3 Masters being built that year alone.

The Aeronca C-3 was also built in England as the Aeronca 100 by Light Aircraft, Ltd. This firm, under the new name Aeronautical Corporation of Great Britain, Ltd., built twenty-four Aeroncas before halting production due to a lack of sales.

The British-built Aeronca was virtually identical to its American cousin with the exception of having more conventional fabric-covered ailerons instead of metal ones. As with the Ohio-built aircraft, the ailerons were interchangeable right to left.

With an E-113C engine of 40 hp, the 1935 PC-3 performed well on Edo floats.

The Aeronca 100 originally was supposed to make use of the latest version of the C-3 engine, the 40-hp Aeronca E-113C being introduced on American aircraft in 1936. However, this engine had single rather than dual ignition and therefore did not meet British airworthiness standards. A license-built version of the E-113C with dual ignition was therefore built by J. A. Prestwick, Ltd., and was designated the J.A.P. Model J99.

That Aeronca in the United States did not incorporate dual ignition in their engines until 1937 was a matter of economics. Two spark plugs in each cylinder running off independent magneto generators meant greater cost for a marginal increase in safety. A more important area for improvement was thought to be the crankshaft, and this component was made considerably stronger in the E-113B of 1934.

Another change in the appearance of the C-3 was accomplished in 1934 when yet another type of landing gear was fitted. The tripodal gear had at first relied solely on the fat Goodyear "airwheels" for shock absorption until complaints of porpoising during rough landings led to the incorporation of an oleo strut. However, this system was bulky and produced excessive aerodynamic drag. The final landing gear consisted of a single cantilever strut with shock dampening housed internally in the fuselage.

The later development of the early Aeroncas was due to Roger E. Schlemmer, who had replaced Robert B. Galloway as chief engineer for Aeronca. Other company officers in 1935 were Taylor Stanley, president; H. V. Fetick, vice president; Robert A. Taft, former senator and son of the twenty-seventh president of the United States, secretary; and Walter Draper, treasurer.

Jean Roche himself parted company with Aeronca in the early 1930s as the major developmental work was past, and he felt a lack of challenge in overseeing the production of his plane. Moreover, the growing size and complexity of the Aeroncas, although more to the public taste, took them ever further from his idea of a minimal light airplane suitable only for sport. From McCook Field, Roche moved to the National Advisory Committee for Aeronautics in Langley, Virginia. There his skills as a developmental engineer were well utilized.

Production of the successful C-3 was terminated late in 1937, not because it showed any drop in popularity but because it no longer met airworthiness requirements for production in this country. The C-3 was loaded with now illegal features such as 3/32-inch control wires, external wire bracing instead of strut bracing, fabric right up to the firewall, a single-ignition engine, and no requirement for an airspeed indicator.

In 1935, Aeronca introduced a low-wing aircraft with the designation C-70 or C-85, depending on whether it was fitted with the 70-hp or 85-hp LeBlond radial engine. With side-by-side seating for two in a completely enclosed cabin with excellent visibility, the C-85 offered a top speed of 120 mph. Overall dimensions were similar to the C-3 and the wingspan was the same at 36 feet, but the gross weight was half again higher at 1,500 pounds. A more powerful version with a 90-hp Warner Junior engine was also offered.

The true successor to the C-2 and C-3 series light planes, however, was the Model K. Introduced in 1937, it was a considerably more modern and pleasing design by all accounts. Powered by a dual-ignition Aeronca E-113C engine, the Aeronca K sat higher off the ground and had a strut-braced wing over an enclosed cockpit with side-by-side seating for two.

While the Model K did away with the comical lines usually associated with the Aeronca name, there were still plenty of "flying bathtubs" around. A grandfather clause in the new rulings allowed all aircraft built before the date of new regulations to continue flying.

Meanwhile, many privately owned Aeronca C-2 and C-3 aircraft were setting an impressive number of records in the Class "C"—4th Category, a division established by the National Aeronautic Association (N.A.A.) especially for light airplanes. Aeroncas seemed to bring out the sporting side of their owners, and the possibilities suggested themselves for altitude, distance, and speed records.

Twelve out of thirty-three Official World's Records recognized by the N.A.A. in 1936 were won by Aeronca C-2s and C-3s, and by the time production of the C-3 ended the following year the figure was up to nineteen. Forest M. Johnston and his C-2 NC-568V held a number of them, predictably enough, including altitude records for both light land and sea planes.

Stanley C. "Jiggs" Huffman flew a C-2 with extra fuel from Lunken Airport to Roosevelt Field on Long Island, New York, with one quick stop for refueling in New Jersey due to unexpected headwinds. Total time for the flight was 10 hours 10 minutes, 9 hours 40 minutes of it in the air, and 30 gallons of gasoline were used to cover the 570-mile distance. Upon landing, Huffman stated that the total cost of the flight had been $9.70, and proclaimed the C-2 the "smallest practical plane in the world."

Huffman had learned to fly in the Army Air Service in 1921, and had worked for Embry–Riddle at Lunken Airport before going into business for himself at the same field. In June 1930, he became involved in another record-setting attempt that became a comedy of errors. With Embry–Riddle backing him, Huffman was to make a try for the solo endurance record in the light-plane class.

The existing record was held by Vern Speich of Long Beach, California.

The Aeronca C-3 was license-built in Great Britain as the Aeronca 100. This example was photographed in Rotterdam, The Netherlands, where it was taken and demonstrated in the hopes of finding a European market.

The Aeronca E-113C engine.

An Aeronca PC-3 flies with a Berliner-Joyce OJ off its wingtip. (Courtesy Peter M. Bowers)

This Aeronca C-2 was sold in Great Britain.

The C-3, like the C-2, was built with a tail skid rather than a wheel as it was designed to be flown off of grass fields rather than hard-surface runways. (Courtesy Kenneth R. Unger)

The standard paint scheme for the C-3 of the mid to late thirties was yellow with a black stripe, or all yellow. This view indicates why the nickname of "flying bathtub" stuck so tenaciously. (Courtesy Peter M. Bowers)

This Aeronca C-3 was used to test an experimental propeller in 1943. The propeller, developed by a Hamilton Standard engineer on his own, was designed to adjust itself in flight to the most efficient pitch. (Courtesy Kenneth R. Unger)

The Aeronca C-2 and C-3 had large, crimped metal ailerons. An unusual feature was that they were interchangeable between right and left.

This brightly painted C-3 was used in airshows.

Another post-World War II C-3 with a bright paint scheme.

Designer Jean A. Roche's belief in simplicity of structure is well illustrated in the Spartan Glider which he designed in 1930. (Aeronca, Inc.)

Ken Unger, American ace who won eighteen victories while flying with the Royal Flying Corps in World War I, poses playfully with a part of the fuselage of a C-3 he plans to rebuild. In the hangar behind him is a complete C-3. This photograph was taken shortly before World War II at Hadley Field, New Brunswick, New Jersey. (Courtesy Edward W. Lawler)

A view of the Aeronca factory in the mid 1930s.

A view from another angle reveals the sales office and completed C-3 aircraft.

Interior view of the Aeronca factory taken in 1936 reveals a low-wing Aeronca C-70 at left and a C-3 at right.

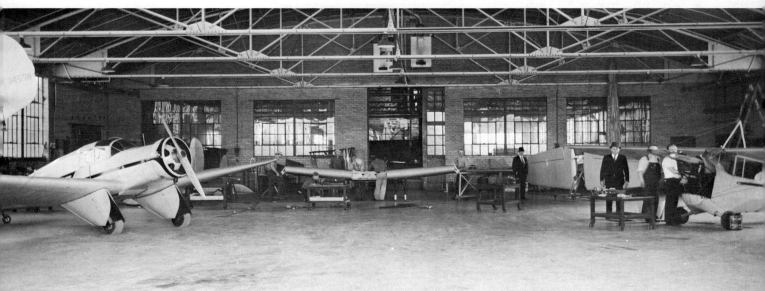

Aeronca Model K.

Stanley C. "Jiggs" Huffman.

Sixteen-year-old Paul Clough in the cockpit of the Aeronca C-2 which he flew to an altitude record.

This unusual photograph shows a restored Aeronca C-2 owned by airline pilot Robert Cansdale flying with the Aeronca C-3 owned by Peter M. Bowers. (Courtesy Peter M. Bowers)

Kenneth W. Scholter set an altitude record of 17,467 feet on April 12, 1931.

Edward W. Stitt and his distance record-setting Aeronca.

Aeronca C-2 NC655W, in which Edward W. Stitt set two light plane distance records.

Robert Bryant and his Aeronca C-3. Bryant, with metal ballast in the second seat to simulate the weight of a passenger, flew from South Carolina to Florida for a distance record of 635 miles.

In late 1935, Benjamin King attained an altitude of 16,000 feet for a new seaplane altitude record.

Speich had remained aloft a bit over 38 hours; if Huffman could fly over Lunken for 40, he would not only break this record but he would also set another for distance covered by a light airplane in one flight. That record was 321.3 miles and had been set two years earlier in Budapest—in this flight, the C-2 would fly more than 2,000 miles!

In addition to the regular 8 gallons of gas, "Jiggs" would have an extra 70 in special tanks to bring the total weight of plane and pilot to 1,062 pounds. A sealed barograph would be carried, and the flight was to be sanctioned by the N.A.A.

The latter gave a full go-ahead for the attempt until they realized at the last minute that this was to be a solo attempt. They then withdrew their sanction, stating in a telegram that the "contest committee does not look with favor on such flights as test is only one of physical endurance of pilot and not of plane."

Not to be daunted, the Embry-Riddle team decided to try instead for a new world's altitude record. For this attempt, the N.A.A. provided a barograph to record the altitude attained on a paper strip.

At 5:32 P.M. on August 3, 1930, Huffman took off and climbed to approximately 16,000 feet before again landing at Lunken at 7:30. The sealed barograph's paper recording strip was sent off to the N.A.A. in Washington right away and confirmation of the record was confidently awaited.

When word came, it was not what had been expected. The N.A.A. stated that no record had been attained as the Aeronca had only climbed to 11,302 feet. In addition, they stated that even had a higher altitude been reached, no record could be recognized as the barograph had not been properly handled.

The instructions for the barograph had been scanty at best, but even so Embry-Riddle felt they had done their part. In frustration, they turned the problem over to Aeronca, their neighbor at Lunken Airport and the builder of the plane. Aeronca examined the barograph, as yet not returned to Washington, and realized it was totally unsuitable for the use to which it had been put. In a letter to the N.A.A., Aeronca asked "how in the world anybody could have obtained official recognition for a new record when that particular barograph had only a range of 11,302 feet?"

If a contrite apology and a better barograph were expected, they never arrived. "Jiggs" Huffman and Embry-Riddle gave up the idea with regret. On April 12, 1931, Kenneth W. Scholter took an Aeronca up to 17,467 feet at Detroit, Michigan, and set the official record Huffman and Embry-Riddle had wanted. Meanwhile, unofficial reports of C-2s attaining 20,000 were not uncommon.

Edward W. Stitt flew his Aeronca C-2 NC-655W, serial #35, from Toledo, Ohio, to Lawrenceville, Virginia, for a distance of 529 miles, a record which he later broke. On July 31, 1937, Stitt flew from Columbus, Ohio, to an alfalfa field just west of Des Moines, Iowa, for a new total distance of 584 miles.

Another distance record was set by Robert Bryant in Aeronca C-3 NR-14657. According to N.A.A. rules, all seats in the record-setting aircraft must be filled. Therefore, three iron weights were placed beside Bryant as ballast. Bryant flew from the Rock Hill Airport in Rock Hill, South Carolina, to the Pan American Airport in Miami, Florida, for a total distance of 635 miles.

These and similar flights served to demonstrate the abilities of the light plane. It was a serious transportation machine, not a toy to be dismissed casually. Moreover, such record flights worked a gradual change upon the thinking of the general public. That such flights could safely be made served to put aside a portion of the mistrust often evinced by the airplane. It is

The Aeronca PC-3 was a very attractive aircraft that offered real utility to its owner.

The practicality and ruggedness of the PC-3 made it the ideal airplane for transporting people or small cargoes to areas accessible only by water.

An Aeronca C-2 shows off its clean lines. Grass fields and open sky were its natural elements.

interesting to note that a number of the records set by the single and two-seat Aeroncas are still impressive by today's standards for such low-powered planes.

Perhaps the greatest tribute to Aeronca came when one English owner flew his British-built Aeronca 100 on a 10,000-mile excursion from London to Cape Town, South Africa. During the entire trip, no trouble was experienced with either the plane or its engine.

But the true significance of the first Aeroncas lies not in the records they set; rather, it stems from the role they played in the beginnings of general aviation in the United States. Although poorly defined or relatively unknown in the minds of most people, general aviation is today the world's largest air carrier, and transports more people between United States cities than the nation's three largest airlines. Only 500 airports are served by airlines in the United States whereas 13,000 are used by sport, family, and business aircraft. Corporate jets, forest-fire water bombers, flying ambulances, traffic and patrol helicopters, and small cargo planes are just a few examples of the roles played by general aviation aircraft.

With their dragging bellies and pointed noses, wire-braced wings and squat appearance, the first Aeroncas could never be called beautiful. The C-2's comical lines were if anything accentuated in the pot-bellied C-3, but whatever the "flying bathtubs" may have lacked in visual impact was more than made up for by the floodgate they opened. They allowed the average man to indulge in a dream of flight, and led a revolution in aviation that affects the lives of every person in the United States, however indirectly.

And it all started with the Aeronca C-2. The first example of this light plane is preserved in the National Air and Space Museum of the Smithsonian Institution. There it will remain so that future generations may see America's first successful light airplane.

The National Air and Space Museum's Aeronca C-2, the first production Aeronca, is seen here shortly after being built in 1929. (Peter M. Bowers)

III

Restoration of the First Production Aeronca

The National Air and Space Museum's Aeronca C-2 was flown privately by no less than thirteen separate owners before being retired in 1940. At that time, Aeronca became aware of its existence and reacquired it for display at their new plant in Middletown, Ohio. A brief history of its active life is included in Appendix II.

The little Aeronca, N626N, was in surprisingly good condition although it had been rather extensively modified during its career so that it resembled a later version of the C-2. The nose had been changed, the vertical stabilizer and rudder were from a C-3, and a tripodal landing gear with balloon tires now replaced the earlier cart-type wheels and single axle through the body of the plane.

In 1946, the National Air Museum was established under the Smithsonian Institution by an act of Congress, and the first meeting of the new museum's advisory board was held in December of that year. The board consisted of Maj. Gen. E. M. Powers, representing Gen. Carl Spaatz, Chief of Air Forces; Rear Adm. H. B. Sallada, representing Adm. Chester Nimitz, Chief of Naval Operations; Messrs. William B. Stout and Grover Loening, civilians appointed by President Truman; and as chairman, Alexander Wetmore, sixth Secretary of the Smithsonian Institution.

The first official act of the board was to put out a call to all aircraft manufacturers and related groups for items worthy of inclusion in the new National Aeronautical Collection. John W. Friedlander, president of the Aeronca Aircraft Corporation, responded with news of the availability of the "first Aeronca airplane ever made," and suggested that the National Air Museum might like to have it.

In a letter dated February 20, 1947, Friedlander went on to state, "We have this ship in our hangar here in Middletown and it is available for your inspection any time. Needless to say that if you consider this plane to be of historical significance for your purpose, we will be more than glad to put the airplane in excellent shape."

Upon acceptance of the Aeronca C-2, there immediately arose the problem of storage space as the National Air Museum existed more in name than in fact. On Gen. Henry H. "Hap" Arnold's order, sixty World War II aircraft of all nationalities were gathered together for the new museum at a wartime Douglas DC-4 assembly plant located at Park Ridge, Illinois, and it was decided that the Aeronca would join them. It was shipped, disassembled and crated, and arrived to find itself in the illustrious company of such types as

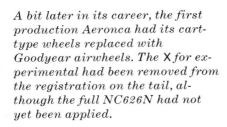

A bit later in its career, the first production Aeronca had its cart-type wheels replaced with Goodyear airwheels. The X for experimental had been removed from the registration on the tail, although the full NC626N had not yet been applied.

the Mitsubishi A6M "Zero," Messerschmitt Bf-109, and Boeing B-17 Flying Fortress. With the advent of the Korean War, the plant was again needed at Park Ridge and these airplanes were moved to the National Air Museum's storage facility at Silver Hill, Maryland.

Aircraft restoration is a slow and costly process, and with so many planes requiring attention it was to be many years before the Aeronca's turn came. The tremendous growth of general aviation in the United States gave impetus to this project, however, and in late 1973 the first steps were taken. At this time, it was discovered that the plane was technically on long-term loan and not owned by the museum, and Aeronca responded with habitual courtesy to transfer ownership to the National Air and Space Museum, as it is now known.

Meanwhile, a local chapter of the Antique Airplane Association had just completed restoration of the museum's Curtiss F9C Sparrowhawk on a voluntary basis, and was casting about for another project of a smaller scale. They took up the Aeronca C-2 until they were forced to stop as all shop activity at that time had to be devoted exclusively to preparing aircraft for the move into the new National Air and Space Museum building. The C-2 was not included in these plans and had to be set aside for the time being.

Following the opening of the museum on July 1, 1976, new plans for aircraft restoration were drawn up and unfinished projects were given high priority. Research was completed and exacting guidelines were drawn up by museum curators for putting the Aeronca back in the configuration and markings it wore when it emerged from the factory in 1929. Museum craftsmen undertook the project and work started on September 14, 1976.

Before the restoration, a bit more of the earliest Aeronca's history came to light when Walter Roderick, chief of restoration at the Silver Hill Facility, discovered that lettering had been painted over on the fuselage of the C-2. He exposed and repainted it so that the information would not be lost.

"The first Aeronca," it read, "America's first light plane. Built . . . June 27, 1929. Flown . . . 3571 hours. Flown . . . 352,320 miles." These figures

Although scarcely recognizable as the plane in the previous photograph, this is how NC626N appeared in 1940.

cannot be verified, but they suggest that this airplane was flown extensively in the decade before it was retired.

The restoration team consisted of John Cusick, Carroll Dorsey, George Genotti, and Patricia Williams. Walter Roderick also lent a hand on occasion, as did others among the skilled Silver Hill staff. The first step was to remove the aged fabric from the plane. This fabric was put aside to serve as a pattern in applying new fabric at the end of the restoration project. The original stitching was carefully noted so that it could be exactly duplicated.

The fuselage needed extensive work as the bottom longeron was bent at the tail skid attachment, a common occurrence among "tail draggers" or conventional gear aircraft. In addition, the wood used in fairing strips was warped, the throttle pivot was corroded so that the throttle was frozen in place, the bumper pad on the rear of the fuel tank was not installed, the seat cushions were rotted, the windshield had deteriorated and was only partly installed, and the plywood floorboards were rotted and saturated with oil.

The list of complaints grew as each piece was checked and noted. The stick grip had deteriorated as had the painted surface of the instrument panel, the tires were corroded, the aileron cables in the fuselage were rusted, the rear of the cockpit had been modified with a tube brace and shelf, and the seat belt and attachment clamps were missing.

Of the instruments, the altimeter glass was broken and the glass gasket had dried out, and the oil temperature and tachometer instruments were

Among the modifications made to the first Aeronca was the addition of tripodal landing gear, illustrated to good advantage in this view of the C-2 at Silver Hill, Maryland.

The first step in the restoration process was to disassemble the plane and remove the old covering of fabric.

This side view shows the C-3 vertical stablizer and rudder which was fitted to this C-2.

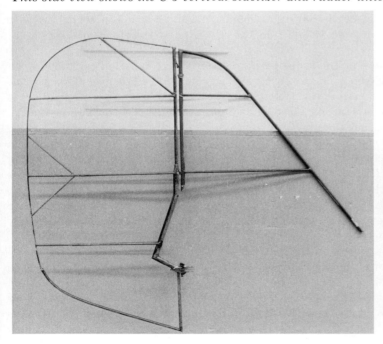

The verticle stabilizer and rudder removed from the C-2.

The new, original-type stabilizer and rudder built up by Silver Hill craftsmen.

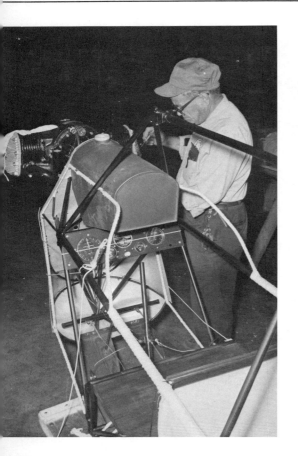

George Genotti, museum sheet-metal specialist, makes some finishing touches on the installation of the Aeronca E-107 engine. Note that the C-2 was built with a rudder bar rather than pedals. (Courtesy Robert C. Mikesh)

not of the types originally installed in 1929. Locating replacement instruments was just one task among many that faced the curators and restorers.

The stripped fuselage was sandblasted using glassbeads instead of sand for a finer finish. The control stick and rudder bar were also given this treatment to remove the corrosion. The tubular steel frame was then inspected for cracks, rusted areas, and other damage before being covered with zinc chromate primer and epoxy paint. Colors used were gray from the tail to the cockpit and black from there to the firewall.

New floorboards were manufactured locally, painted black, and installed when dry. Control cables were chemically cleaned and coated to prevent future corrosion before being reinstalled. Instruments and nameplate were removed from the panel, and it was stripped of paint while the individual instruments were being worked on. In the Aeronca C-2, the panel contains an absolute minimum of instruments—just oil temperature, oil pressure, nonsensitive altimeter, and tachometer—and is mounted below the rear end of the 8-gallon fuel tank which projects into the cockpit.

An interesting part of the treatment of the fuselage framework called for the drilling of a drain hole at a low point in the steel tubing. This hole was capped before hot linseed oil was poured into another hole at the highest point of the forward end of the structure. The course of the oil could be followed by feeling the heat with the fingertips, and when the entire structure was filled it was allowed to sit for five minutes before the cap was removed. When the oil had drained away, the holes were sealed and the restorers were left with a frame which would not corrode internally.

The wings needed considerable attention. Many ribs were damaged at their thin trailing edges, and the top camber of several had been flattened by time. The contraction of the fabric over the years had also scalloped the trailing wing edge between ribs, much as in a SPAD fighter of World War I, and there was damage to the leading edge as well. Wing fittings, aileron cables, and shackles were rusted, and little varnish remained on the wing spars.

The fabric on all the tail surfaces had rotted, and the control cables were extensively corroded. In addition, the rear edges of the elevators and rudder were rusty.

At the other end, the propeller laminations were splitting and one propeller tip was broken, held on only by the metal leading-edge piece. This was the one part of the Aeronca C-2 that was left to an outside concern to rebuild, and was sent to the Sensenich Propeller Company of Lancaster, Pennsylvania.

The engine had also suffered the ravages of time. The Aeronca E-107 engine was corroded on the back side, and there was considerable corrosion also on the accessory case, parting surfaces, and firewall. Neither spark plug was installed, the fuel strainer's glass bowl was missing, and exhaust pipes were rusty, there was corrosion on both cylinder heads, and the oil filler cap gasket was missing. In addition, the top and bottom firewall sections were cracked in several places and the metal was creased, dented, and actually chafed through in spots.

The engine was completely disassembled, all parts being cleaned to remove paint or carbon. This E-107 is a very early model as evidenced by the fact that it has no crankcase fins, and is therefore in all probability the same engine with which the Aeronca left the factory. To museum curators whose primary concern is accuracy, this fact meant that a major difficulty was avoided; finding such an engine would have been extremely difficult. Today, the small engine once again appears new and is coated internally with preservatives.

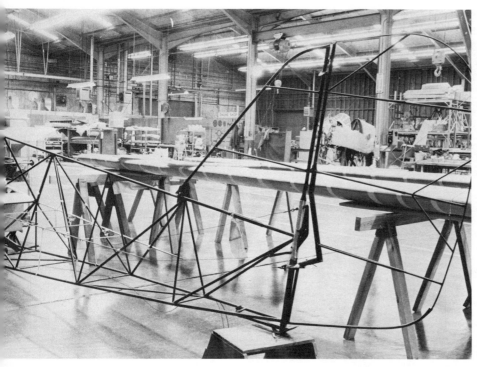

The triangular steel-tube fuselage can be seen to good advantage in this photograph. In the background, a P-51C is being restored. (Courtesy Robert C. Mikesh)

John Cusick joins the new rudder to the fuselage of the C-2.

The Aeronca fuselage, masked and ready for the first application of paint.

The engine installation features an updraft carburetor and single ignition. There is no electrical system or starter, and the engine was started simply by swinging the propeller.

The fuselage was painted yellow and orange, as it appeared when it left the Aeronca factory in 1929. Wheels of the original type have been reinstalled.

The completed fuselage awaits the wing, engine cowling and propeller.

The wing required extensive work. Original wood was used wherever possible, and new pieces were marked as replacements according to museum policy.

The bent fuselage longerons were straightened in a specially built jig before John Cusick and his group began the long process of substituting original-type tail surfaces for those on the plane. This involved more than just replacing the rudder, as the vertical stabilizer had been modified as well. These sections were built up from scratch using tubing of the original diameter and following drawings provided by Aeronca especially for this project.

New Grade-A cotton fabric was then applied to the fuselage and the original stitching was carefully duplicated. The fabric was then dampened with water to remove the wrinkles and a coat of fungicidal dope was applied. This surface was then sanded lightly with fine sandpaper, and additional coats and colors were applied.

The fuel tank, which sits exposed between the cockpit and the engine cowling, was painted the same orange color as that part of the fuselage. It was then installed and all fuel lines were connected.

A new windshield was cut out, using the old one as a pattern, and new windows were put in place above the pilot's seat. Cushions copied from the

John Cusick attaches new Grade-A cotton fabric to the C-2 wing.

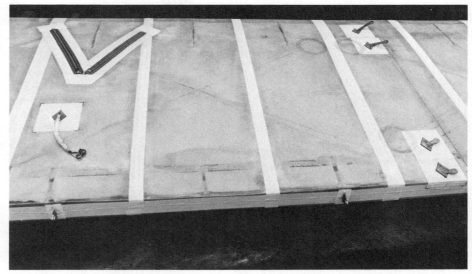
After being covered with fabric, reinforcing tape is applied over the ribs.

Fungicide is applied to the fabric before it is covered with aircraft dope.

Locating original-type wheels was a major stumbling block in the restoration effort. The solution came when it was learned that carts used for exercising harness race horses use similar wheels, and a set was obtained through a local saddlery shop.

The rest of the landing gear assembly, including the cover plate and axle, was manufactured by museum craftsmen according to original factory drawings.

originals were put in place and the Aeronca C-2 fuselage began to look more businesslike.

The freshly overhauled engine was installed and all controls were connected, as were the fuel and oil lines and the tachometer cable. The reworked engine cowling and exhaust stack were fitted, the carburetor heat box was attached to the carburetor, and the remanufactured wooden propeller was bolted on to complete the fuselage front end.

In the meantime, the wings were being repaired on a nearby table. Old varnish was removed and the entire wing, minus the fabric, was washed down with acetone. Previous repair work was undone and all ribs were inspected for dry rot. Where necessary, wood was cut out and replaced. Even where replacement was necessary, 1/16-inch gussets were glued on both sides of the rib for reinforcement.

Ribs that had flattened out were steamed and formed back into shape, the cap strip being clamped into place until dry, and stiffeners were glued and nailed as required. The metal leading edges were also removed in order to facilitate repair. The crimped metal ailerons were in good shape and required only cleaning and treating against corrosion. A new trailing edge was manufactured and installed in the same manner as had been done at Lunken Airport near Cincinnati almost a half century before.

Rusty metal fittings were sandblasted with glassbeads, and cadmium-plated fittings were replated before being reinstalled. The entire wing structure was revarnished before new fabric was applied. Reinforcing tape was applied over every rib from the leading to the trailing edges, and the stitching was done. The dope used was nontautening so as not to pull the rather fragile wing structure out of alignment.

The next major challenge was the landing gear. The entire split-axle, tripodal assembly was discarded as it was not original to this airplane. Harvey Napier, the machinist for the restoration shop, manufactured from scratch an entire landing-gear assembly of the proper type from original factory drawings. It incorporated a through-axle with shock-cord dampening.

Wheels were quite a problem, for all that was known about them was the tire size and what could be seen in pictures. A quick check revealed that these were not tires that could be easily obtained at the nearby bicycle shop. Inquiries were then made with motorcycle dealers and tire distributors in various parts of the United States. Calls were also made to Canada, and sources there in turn contacted England, but with no success. The hope had been that if a tire of the right size could be located, finding what wheel it fitted would not be a problem.

The problem was seemingly without a solution, and it was beginning to appear that both wheel and tire had long since been out of production. Then, a sudden inspiration provided the answer—southwestern Ohio, where the Aeronca had been produced, was horse country. Might the wheels be associated with horse-cart racing? A call to a harness and saddlery shop less than a mile from Silver Hill confirmed that such wheels were used on carts for exercising harness race horses. The wheels were obtained and installed, their spokes covered as on the original, and the original type of landing gear was at last achieved.

Care of the most exacting and painstaking nature was taken by National Air and Space Museum curators to ascertain the exact colors and markings the restored aircraft should bear. Aeronca employees, aviation enthusiasts and historians—even designer Jean Roche himself—contributed to this effort. One example of the difficulties this task entailed was accurately repro-

Accurately reproducing the Aeronca "Wing" emblem on the tail of the C-2 was a big problem.

Robert C. Mikesh, curator in charge of the restoration of the Aeronca C-2, examines the freshly restored airplane.

Looking as sleek as the day it rolled out of the Aeronca factory a half century before, the National Air and Space Museum's C-2 appears once again ready to take to the air.

ducing the colorful Aeronca "wing" emblem on the tail of the C-2. Several versions of various colors and sizes were used in the early years of the company's history, and determining the authentic configuration required a considerable effort by the museum curators and John Houser, service engineer of Aeronca, Inc.

After the little C-2 was completed, it was given a 10-day rest while the dope seasoned. Then all surfaces were polished with a good grade of wax, and the plane was buffed to a shine. On July 14, 1977, the completely restored Aeronca was wheeled out into the summer sunshine to be photographed. Total manhours spent on the restoration project amounted to 2,234, and the total number of manhours spent actually working on the airframe and engine reached 1,806.

The National Air and Space Museum's Aeronca C-2 is restored inside and out, and is as authentic as possible. As is the policy of the Smithsonian Institution, where original parts are replaced, the newer pieces are marked to reflect the fact. These aircraft restorations are more than just cosmetic; they are designed to insure that the airplane will remain in good condition for many years to come. A question often asked is whether or not the restored aircraft are flyable. The answer is that little work would be required to bring them up to flying status, but they will in all probability never fly again. Rather, they will be tangible reminders of the history and development of aviation.

The first Aeronca ever built will no longer take to the air. But it may evoke smiles and fond memories from those who see it, and it may eventually surmount the obscurity into which it has fallen. Gleaming resplendently in its yellow and orange paint, it should long serve to focus attention on the beginnings of general aviation in the United States.

A. Specifications and Performance, Aeronca C-2

WINGSPAN	10.98 m (36 ft)
WING AREA	13.2 sq m (142.2 sq ft)
WING LOADING	24 kg/sq m (4.92 lb/sq ft)
LENGTH	6.10 m (20 ft)
HEIGHT	2.28 m (7 ft 6 in)
WEIGHT	Empty 184 kg (406 lb)
	Gross 318 kg (700 lb)
TOP SPEED	128 km/h (80 mph)
CRUISE SPEED	104 km/h (65 mph)
STALL SPEED	50 km/h (31 mph)
FUEL CAPACITY	27.3 l (8 U.S. gal)
RANGE	384 km (240 mi)
SERVICE CEILING	5,032 m (16,500 ft)
ENGINE	Aeronca E-107A, 2-cylinder, horizontally opposed, single ignition, 26 hp
MANUFACTURER	Aeronautical Corporation of America, Inc., Cincinnati, Ohio

B. History of the First Production Aeronca, 1930-1940

OCTOBER 4, 1930	Herman and Macy Teetor of Hagerstown, Indiana, buy Aeronca C-2 NX626N direct from factory. Aircraft registration changed to NC626N. The plane is based at Issoudon Field, Hudson, Ohio.
JUNE 30, 1931	NC626N sold to Carl K. Wollam of Hudson, Ohio. Plane still based at Issoudon Field.
AUGUST 16, 1931	Wollam sells the earliest Aeronca to William T. Koldt of Hudson, Ohio.
NOVEMBER 23, 1931	NC626N is purchased by Paul E. Wein of Akron, Ohio. The plane is based at Mid-City Airport, Hudson, Ohio.
MARCH 26, 1932	Wein sells the Aeronca C-2 to Miss Gertrude A. Fravel of Lakewood, Ohio. Miss Fravel bases the plane at the Cleveland Municipal Airport, Cleveland, Ohio. It is thoroughly inspected in Cleveland after 77 hrs and 25 min of flight time, and given a clean bill of health on June 29, 1932.
OCTOBER 15, 1932	Miss Fravel sells NC626N to Robert M. Stevenson of Palmyra, Missouri. Stevenson bases the C-2 at Long Airport, Hannibal, Missouri.
MAY 24, 1935	Stevenson sells the Aeronca to Elmer C. Long, owner of Long Airport.
SEPTEMBER 10, 1935	Long sells the first production Aeronca to Dan R. Stewart, who bases his plane at Long Airport.
APRIL 13, 1936	Stewart sells NC626N back to Elmer Long.
DECEMBER 10, 1937	Long sells the Aeronca to C. J. Tolliver of Hannibal, Missouri.
MARCH 12, 1938	Tolliver sells NC626N to Cloyd Schumm and Lawrence McKay, both of Hannibal, Missouri. Plane still based at Long Airport, and is modified by Long to incorporate a new vertical stabilizer and rudder of the type used on early model Aeronca C-3 aircraft. Drawings for this project were supplied to Long by Aeronca in late October 1938. The modified aircraft is test flown by Long on April 3, 1939, and put in spins to the left and right before observing aeronautical inspectors of the Civil Aeronautics Authority.
AUGUST 19, 1939	Schumm sells out his half interest to his partner, Lawrence McKay, leaving McKay the sole owner of the aircraft.
MARCH 22, 1940	McKay sells the C-2 to Elmer Long. It has been based at Long's Airport since 1932, and this becomes the third time it is owned by Long.
JUNE 10, 1940	Long sells NC626N back to Aeronca, who is anxious to reacquire the plane for display at its new factory in Middletown, Ohio. Eight years later, Aeronca, Inc., donates the historic aircraft to the National Air Museum of the Smithsonian Institution.